Y0-CDB-896

Pest Control

You Can Live With

**Safe and Effective Ways
To Get Rid of
Common Household Pests**

by Debra Graff

Earth Stewardship Press
Sterling, VA

LIBRARY
ST. LOUIS COMMUNITY COLLEGE
AT FLORISSANT VALLEY

Notice

The information in this book has been carefully researched and is presented to provide basic information on some alternatives to chemical pesticides for home pest control. The author has attempted to warn the reader of the dangers of using various methods. However, the reader should realize that, as with all chemicals, dangers may exist that have not been addressed in this book. In order to insure the safe use of any pest control method, the user must read and closely follow all directions on pest control product labels and consult a pest control expert if necessary. Earth Stewardship Press and the author assume no responsibility for any injuries suffered, or other damages or losses incurred during, or as a result of, following the information provided in this publication.

Copyright © 1990 by Debra Graff

All rights reserved. Printed in the United States of America. No part of this publication may be reproduced or distributed in any form or by any means, without written permission of the publisher, except for the inclusion of brief quotations in critical reviews or articles.

First Edition 10 9 8 7 6 5 4 3 2 1

Publisher's Cataloging in Publication Data

Graff, Debra A.
  Pest Control You Can Live With: Safe and Effective
    Ways To Get Rid of Common Household Pests

1. Household pests - Control. I. Title.
Library of Congress Card Catalog Number 90-061406
ISBN 0-9626191-9-1

Published by: Earth Stewardship Press
                P.O. Box 1316
                Sterling, VA 22170

For book ordering information, see page 80.

Printed on recycled paper.

Dedicated to our planet, Earth.

Acknowledgments

I would like to thank my parents, Bernard and Pauline Graff, and my sister, Brenda Graff. Without their unending support and assistance, this book would never have been written. I would also like to thank my friend, Linda Titolo, for her generous and positive feedback throughout this project.

Contents

Introduction

Ten years ago, I lived in an apartment with two cats, one roommate, and about ten million fleas. The fleas ignored my rommmate, and chose to feast on my legs. I had dozens of bites on any one day, and I itched terribly for weeks after each one. Although I was studying for a degree in organic agriculture, I was very ignorant of the life cycle of fleas, and of any safe and effective methods to control them.

There weren't too many options in 1980. I didn't know about flea combs or the value of frequent and thorough vacuuming. Herbal flea collars didn't help, and my cats hated their smell. I bought a bottle of liquid pyrethrum (the kind used in gardens), and proceeded to frantically spray my apartment with it. The result was a white film covering all the surfaces I sprayed. Unfortunately, I had found out the hard way that the product was not designed for indoor use. I had less than a week of relief from the fleas before there were suddenly ten million more.

Although I was desperate to get rid of the fleas, I had a strong aversion and concern about using chemical pesticides—especially in my own home. I knew that most of them were forms of nerve poisons, and that nearly 90% of the pesticides on the market had never been thoroughly tested for long-term health problems. Finally, feeling that I had no other option, I resorted to chemical flea collars, which, luckily, brought the fleas under control.

That winter I moved, and had to give away my wonderful,

though troublesome, cats. In the years since, in various homes, I have had to deal with infestations of cockroaches, flies, and even (in one truly gruesome house) slugs that ate the posters hanging on my walls. I realized that organic gardening wasn't enough. I had a driving need to discover safe and effective control methods for pests <u>inside</u> my home.

Fortunately, research in the last decade has greatly widened our options for safe indoor pest control. We now have a wonderful range of products and methods to choose from, and the choices are increasing every year. Although this book does not include all of the options available, I have given instructions for most of the more reliable and effective ones.

I welcome comments from you about your experiences with the various methods and products mentioned in this book, or with any others that you may have tried. (See page 79 for instructions on how to contact me.) Such information could help make the next edition of this book even more useful for everyone.

Good luck with your pest control efforts!

Chapter 1

How to Use this Book

Choosing control methods

A wide variety of pest control methods and products are described in this book. It is not necessary for you to use all of the methods listed in each chapter. Choose the ones that are best suited for your particular situation.

Not all of the control methods and products mentioned in this book are likely to be "miracle cures" for your insect problems. A number of them will be very effective, with a success rate of 90-99%. Some will be less effective, and a few may not work very well for you at all. There are many factors involved in determining how successful a method or product is. Sometimes it may even depend on the weather.

If the first method you use doesn't work very well, experiment. Try a different method, or combine two or three to achieve the best control. Choose the methods and products best suited for your particular situation, and follow the directions in this book and on the product labels carefully.

The very safest methods include the following:

1) Reducing the sources of food, water, and shelter for pests in your home
2) Sealing the insects' entrances into your home

3) Using mechanical control techniques, such as traps, vacuuming, and flea combs

Whenever possible, these methods should be used first. The major exception would be a severe infestation requiring immediate control. In this case, the use of the stronger products mentioned in this book can quickly bring the insect population under control. Once their numbers are brought down to low levels, the safest methods mentioned above may be able to keep them down.

Safety precautions

Always read the product labels very carefully. There are many products on the market that combine the safer and the more dangerous pesticides together.

You should also read completely through this book, and pay close attention to the products that you will be using. Technical boric acid is not the same as medicinal boric acid. Food grade diatomaceous earth is not the same as swimming pool filter diatomaceous earth. Regular soaps and detergents are not the same as Safer® soap products. This book explains which products are safe and effective to use, and which should be avoided.

Follow the simple safety precautions described in this book, and those listed on product labels. Don't let these basic precautions scare you. Most of them are common sense:

1) Keep these products out of the reach of children.

Federal law requires this label on all insecticides, no matter what their degree of safety. This is a good precaution to follow, as children have often injured or killed themselves with items as "safe" as loose buttons, plastic bags, and buckets of water left standing around. Always keep all pesticides in a child-proofed or locked cabinet. Be very careful with any pest control products if you have young children visiting or living in your home.

2) Unless the product label states that it is safe to eat, don't swallow it.

Some of the products mentioned in this book are toxic if swallowed. These products should also not be used on or near your food, or your cooking or eating utensils.

3) Avoid breathing in the dusts and sprays.

Dusts and sprays of any kind can irritate your lungs, or cause other health problems. It is an excellent idea to open windows and doors to provide good ventilation when you are applying any pest control product. (If it is very windy, open them only a little way.) Use a simple disposable dust mask when you are applying any powders.

4) Wash your hands, and, in some cases, change your clothes after applying the product.

The products described in this book are usually quite safe when used properly, but a few of them can dry out or irritate your skin with prolonged exposure. In addition, some people can have either mild or very strong allergic reactions to a product, especially pyrethrum. Always closely follow the directions on the product label.

Equipment

With several of the products mentioned in this book, some simple equipment can make their application both easier and safer.

1) Dust or powder applicator

The proper equipment for this depends upon the size of the area that you will be covering, and where you will be applying it. For small amounts, a plastic squeezable bottle (such as a picnic ketchup bottle) will work well, and it can aim the powder into cracks. There are also commercial rubber bulb dusters and other powder applicators that can aim the powder into cracks. A simple way to apply powder over larger surfaces is to use a flour sifter (it may take a little practice to do it evenly). For large areas outdoors, other applicators, such as the Dustin Mizer Duster, are very helpful.

2) Spray applicators

For small areas, a one quart trigger sprayer will do fine, or even a well-rinsed bottle of Windex. To cover large areas in your home or yard, it is easier to use a one or two gallon pump sprayer.

3) Disposable dust mask

To protect your lungs from irritation when you apply powders, these simple masks are well worth the few cents that they cost. Good ventilation also helps to protect your lungs.

4) Rubber gloves

If you are using a product that can dry out or irritate your skin, or that you may be allergic to, protect your hands by wearing rubber gloves.

To locate these products, check at hardware stores, garden centers, department stores, building supply stores, and the mail-order suppliers listed in the appendix of this book. You may want to take advantage of products that come prepackaged in squeezable dusting containers or in ready-to-use sprayers.

If you are not going to use new equipment, please take the following precaution:

If a duster or sprayer has been used with toxic pesticides previously (even it was was last used over a year ago), it must be washed out very thoroughly. Wash it out completely at least three times, including all of the hoses. Do not let the water touch your skin, and do not drain it where pets or children can get into it. Pour the water onto unpaved ground, away from food plants. Do not put it down the toilet, or into the septic system or public sewer system.

Chapter 2

Fleas

Fleas are probably one of the most annoying and frustrating insect pests in our homes. They are often difficult to control even with powerful chemical pesticides. Many people and pets develop an allergy to flea bites that can cause severe itching and irritation for weeks.

Fleas and their young thrive during warm, humid weather. Given these conditions, they can complete their lifespan within just two to three weeks. The adult female flea can lay nearly one thousand eggs during her lifespan, and the average dog can be a host for several dozen fleas at one time. The many thousands of eggs laid can quickly build up the flea population to massive numbers.

Eggs can be laid when the flea is either on or off the host animal. The eggs then fall off the host to join the others that have accumulated on furniture, rugs, bedding, cracks in the floor, and soil in the yard. When the eggs hatch in a few days, the larvae feed on dust, organic matter, and the feces of digested blood from the adult fleas.

After anywhere from one week to several months, the larva spins a cocoon to pupate. In this form, the flea is well protected from most pesticides. It will rest in the cocoon until the temperature and humidity levels are adequate, and it senses the vibrations or warmth from a potential host. An infested house can lay vacant for over a year and still hatch out hordes of hungry fleas when new residents move in.

CONTROL METHODS

Closely monitoring the flea population on your pet, and in your home and yard, is very important. In doing this, you can get a head start on controlling an increasing flea population, often before you see fleas in your home or flea bites on your legs. It also allows you to evaluate and adjust or change the control methods you are using.

There are two simple methods for monitoring the number of fleas. A flea comb can be used to catch and count the fleas on your pet. (See the section on flea combs on page 18 for directions.) You can evaluate the flea population in your home and yard by wearing knee-high white socks while you walk around the area. The dark colored fleas are clearly visible when they jump onto the socks. When you can see more than a very small number of fleas, you know that you need to reduce their population. Check your socks often during your walk-through so that you know exactly which rooms in the home or areas outside need to be treated.

Controlling the fleas on three different fronts—on your pet, in your home, and in your yard—is usually necessary if you have anything other than a small flea infestation. Treating the fleas on your pet alone will not reduce the problem significantly if there are a large number of fleas in the environment. For best results, always treat your animals, home, and yard at the same time.

There are a wide variety of control methods described in this chapter. Choose the ones that you feel comfortable using, and that are most effective for your particular situation.

On your pet

1) Flea comb

These combs are designed to catch fleas between their teeth as you groom your animal. With practice, most animals come to appreciate being flea groomed. Be gentle, but make sure the comb reaches all the way down to the pet's skin. This may be difficult or impossible with some breeds of dogs with thick coats. With these dogs, you may need to concentrate on the areas where the hair is thinner, such as the stomach.

After each stroke, check the comb for fleas and flick them into a bowl or sink of soapy water. You can also wash off the flea eggs (white specks) or flea feces (dark specks) that it may pick up. Don't ever crush fleas between your fingers as they often carry parasites and disease organisms.

Flea combs are very helpful in monitoring the flea population so that you know when to change or increase the control methods you are using. You can either check the whole animal, or choose an area on your pet that the fleas seem to concentrate on, such as the neck or base of the tail. Use a flea comb on that area every day, and keep track of the number of fleas you are catching. This will tell you when the fleas are increasing or decreasing, often before you see any fleas in your home or flea bites on your family.

If your pet appears to be scratching at fleas when you can't find any, look for dark specks on your animal's skin. Comb some of these specks off and put them on a damp paper towel. If they turn into reddish spots after a few minutes, they are flea feces rich in blood from your pet. This indicates that the fleas are still active. To confirm this, bathe or thoroughly flea comb your animal to remove all of the existing specks. Then check the skin a few days later for new deposits of flea feces.

With milder flea infestations, some people are able to maintain very good control with a thorough, daily flea combing of their pets. Sometimes combing just the cats can also reduce the number of fleas on the dogs. This is because fleas jump on and off all the animals frequently throughout the day, so eventually you are likely to catch them while they are on the cat.

2) Safer® Flea Soap

For dogs with coats too thick to be flea combed, bathing is often the best choice. Depending upon how numerous the fleas are, dogs can be bathed monthly or weekly (or even every two days, in extreme cases). As cats often dislike baths, flea combing is usually the best option for them.

Safer Flea Soap is a very safe and effective product to kill adult fleas on your pets. It will also dislodge ticks, but it may not kill all of them. The active ingredients, potassium salts of fatty acids, are made from naturally occurring compounds often found in everyday foods. This concentrated soap can sting the eyes, mouth, nose, and open wounds, and it should not be swallowed. Following these precautions, wet your animal with warm water and work up a soapy lather with this product over the entire body. Let the lather sit for five minutes, and then rinse the animal thoroughly. Do not use Safer® Insecticidal Soap on your pet. It is a much stronger concentration, and may be too drying to the skin.

3) Citrus shampoo

Another safe product for flea control is an extract from citrus rinds, d-limonene. It will kill both adult fleas and ticks. There are now several shampoos on the market that contain this compound. Always read the labels carefully to be sure that the shampoo does not also contain another pesticide.

While it is possible to produce your own extract from citrus fruit at home, it is not generally advisable. If you make the solution too weak, it will not work well, and if it is made too strong, it may sting or burn the skin of your pet. The commercial products are always at safe and effective concentrations.

If you use d-limonene products frequently on your cats, observe them carefully. A few cats may suffer some mild reactions to the product, including depression, excessive salivation, and general illness. If this occurs, simply switch to a different shampoo, such as Safer Flea Soap.

4) Good health

One of the ways to reduce flea infestations on your pets is to keep your animals in good health. It is well known that parasites, including fleas, are more strongly attracted to animals that are weak and in poor health. Ideally, your pet's lifestyle should include a healthy diet, regular exercise, natural sunlight, frequent grooming, and lots of love and attention. Sometimes an animal which is best suited for a different climate, such as a heavy-coated Husky dog living in the southern states, will have more problems with fleas.

5) Brewer's yeast and garlic

Some people have found that adding brewer's yeast or garlic to their pet's diet has helped to reduce the number of fleas.

Brewer's or nutritional yeast (not the dry yeast to make bread) can be added to your pet's food in fairly small doses (about one-half teaspoonful per day for a small cat, and up to two or three teaspoons a day for larger dogs). If too much yeast is fed to your animal, it can produce painful cramps and gas. For the few animals allergic to the yeast, yeast-free B-complex vitamins are available in health food stores. Ask your veterinarian for a safe suggested daily dose, as it can be dangerous to overdose on any vitamin.

You can add one or two crushed garlic cloves daily to your pet's food, if you don't mind them having mild garlic breath.

There are commercial products available that combine yeast, garlic, and other ingredients into snack-type food that is easy to feed to your animal. One of them is Natural Animal™ Yeast & Garlic Bits. In addition to the yeast and garlic, it is fortified with herbs, vitamins, amino acids and other products to help improve the general health of your pet. If you can't locate a source for it locally, you can find mail-order suppliers listed in the appendix of this book.

6) Herbal repellants

Many people have tried controlling fleas with the use of herbal flea collars, often without much success. The strong odors of these collars can be very irritating to the animal or you. Occasionally an animal may become allergic to one of the ingredients, often pennyroyal. Pennyroyal has also been associated with spontaneous abortions. The concentrated oil of pennyroyal should be avoided by pregnant pets and women.

This is not to say that herbal repellants cannot help with your flea problems. If you would like to use an herbal repellant, first reduce the general flea population in the home and yard by other methods. Choose an herbal powder, shampoo, or spray that can be applied over the whole animal. An herbal collar alone doesn't often affect fleas on the rear half of the animal.

Always follow the directions on the product label, and avoid getting it into your pet's eyes, nose, mouth, or open wounds. If your animal develops an allergy to the product, or finds it too irritating, discontinue use. One pleasant-smelling product that contains an insect-repelling herb is Natural Animal™ Spritz Coat Enhancer Spray. You just spray it on your animal, and brush it in when it is almost dry. Some users of this product have said that it will also kill fleas on contact. There are a wide variety of herbal pet shampoos and powders available. You may find some to be more effective than others.

7) Pyrethrum powder

While pyrethrum powder can be a safe and effective product to kill fleas, it is generally a good idea to first try using another less toxic method of control, such as flea combing or bathing. If you find it difficult to use these methods on your pet, you may find it more convenient to use pyrethrum powder. (Be sure to read the instructions on the safe use of pyrethrum on pages 63-65.)

It is best to powder your pet outside. A large dog may require up to an ounce of the powder. Work the dust into the animal's coat over the whole body, keeping it out of your pet's (and your own) eyes, nose and mouth. Try to avoid breathing in the dust. Your pet may become temporarily uncomfortable, as the pyrethrum initially causes the fleas to become agitated before it kills them.

Fleas that don't receive a fatal dose may become stunned, appearing dead, and later achieve a full recovery. For this reason, it is a good idea to brush the fleas out of the animal's coat after all of the fleas have become stunned. You may want to stand your animal on sheets of newspaper while doing this, and then dispose of or burn the dead and paralyzed fleas.

In the home

1) Restricting your pet

It is nearly impossible for pets that have access to the outdoors to stay 100% free of fleas. You will have to decide for yourself what level of flea infestation is tolerable for your pet and family. A much tighter control of the flea population will be necessary if your pet or a family member has a strong allergy to flea bites. In this situation, it may be a good idea to restrict your pet to either the home or the yard (not both), or to keep the animal out of rooms used by sensitive family members.

2) Thorough, frequent vacuuming

Some people find frequent vacuuming by itself to be extremely effective in reducing the number of fleas in their home. Vacuuming reduces the food sources available to the larvae, and removes a large number of flea eggs, larvae, pupae, and adult fleas. The effectiveness of this method depends upon the degree of infestation, the vacuum cleaner, the type of carpeting, and how well and how often it is done.

If you are planning on using vacuuming as the basis of your flea control program, it must be thorough and frequent. Use the white sock monitoring technique to determine how often that you need to vacuum. You may want to vacuum every two to three days in warm weather. With severe infestations, you may need to do it every day until the flea population has been reduced. Two to three times a month may be often enough during the cooler seasons.

Vacuum everywhere in your home that has a flea problem, especially where your pet spends any amount of time. This includes vacuuming upholstered furniture, under cushions and throw rugs, under furniture, and in cracks around baseboards where dust can collect.

Tightly seal and carefully dispose of the vacuum bag so that the fleas cannot escape to reinfest your home. A cheaper alternative to this would be to suck some Diacide into the vacuum after each cleaning. The combination of diatomaceous earth and pyrethrum in Diacide will kill both the adult fleas, and the larvae that would feed and mature on the organic matter in the bag. (For more information on Diacide, see page 65.)

Your pet's bedding should be washed or thrown into a hot dryer frequently to kill the flea eggs and larvae. If the pet spends much time in a favorite chair, keep a washable or dryable blanket or rug there.

If you are dealing with a large number of fleas, it is a good idea to combine vacuuming with other control measures.

3) Steam cleaning

Heavy flea populations can often be brought quickly under control by steam cleaning the carpets and upholstery, combined with a thorough vacuuming of the rest of the house. Safer Indoor Flea Guard can be used in the cleaning solution.

4) Safer® Indoor Flea Guard™

Made from the same active ingredients as those in Safer Flea Soap (only in a less concentrated form), it is one of the safest fast-acting flea sprays available on the market. It is only effective against adult fleas. Use the white sock monitoring technique to determine when and where to treat. Spray all affected areas thoroughly with the solution and then vacuuum 30 minutes later. Be sure to follow the directions on the product label. The rooms can be occupied immediately after vacuuming. Repeat when necessary during the flea season.

As this product biodegrades very rapidly, it does not offer

long-term control of the fleas. A very effective way to use the Flea Guard is to combine it with other longer-lasting control products, such as diatomaceous earth or methoprene. Spray the area with Flea Guard to kill the adult fleas, and then apply methoprene or diatomaceous earth to provide long-term control of the larvae. You may occasionally need to use Flea Guard once or twice more to kill newly hatched fleas.

5) Diatomaceous earth

Many people have successfully used this product against fleas in their home. However, because this is a finely powdered dust, use it with caution. Remember that it is healthier to use frequent vacuuming to control the fleas than to even slightly increase the dust content in your home. Consider using a different control method if you have children that will be jumping around on the furniture and carpets, and raising up more than the usual quantities of dust.

Diatomaceous earth can provide very good long-term control of fleas in your home. It may take up to two weeks to kill off the adult fleas (which is why it's a good idea to use Flea Guard first), but the powder will continue killing fleas and their larvae for several months or until it is washed away. (For more information on the safe application and use of diatomaceous earth, see pages 57-59.)

First, cover or remove electronic equipment that could be damaged by dust (computers, stereos, etc.). It is easier to apply the powder over large areas by using a duster or a flour sifter. Dust the diatomaceous earth lightly onto the floors, rugs, and upholstery (including under the cushions) in every room where you have a flea problem. Concentrate on the areas where your pet spends the most time. Use a stiff brush or wisk broom and work the powder gently but thoroughly into the upholstery, cracks in the floors, and deep into the carpeting. Then lightly vacuum the surfaces.

When you are done, the powder should be invisible. Test your application by striking a carpet or an upholstered seat. If there is a large cloud of dust, you need to vacuum more thoroughly. If there is no dust cloud at all, you need to reapply the dust and brush it in more thoroughly before vacuuming. Use a stiffer brush if you need to. Some carpets, such as indoor/outdoor carpet, badly matted shag, or an extremely short, tight commercial carpet, do not absorb or hold onto the powder very well. If this is the case, use one of the other suggested flea control methods instead.

Diatomaceous earth should not be used in homes of people with lung problems such as severe asthma, emphysema, lung cancer, or similar illnesses. Because their lungs are already impaired, they may have severe reactions to any slight increase in dust in the air. Another method, such as frequent vacuuming or methoprene, should be used to control the fleas in these homes.

6) Methoprene

If you are not able to use diatomaceous earth in your home, and the other suggested methods of control have not been effective, you may want to consider using methoprene (Precor™). It is an insect growth regulator that prevents juvenile fleas from maturing into adults. Methoprene is completely biodegradable, and it is less toxic to mammals than pyrethrin. Insect growth regulators mimic the natural hormones that are unique to insects and related animals—those controlling molting and metamorphosis. Most other pesticides in common use today are far more hazardous to humans because they are poisons that attack the nervous system in both insects and mammals, including people.

Methoprene is frequently combined with other more toxic pesticides, so read the labels carefully. If you have problems finding a local source of a product with methoprene alone, check with wholesale suppliers of pest control products, or contact the Natural Gardening Research Center listed in the appendix of this book.

Precor will not affect adult fleas. Used by itself, it would take many weeks to see any effect on the flea population. Because of this, the adult fleas should first be controlled by other suggested methods, such as steam cleaning or the use of Safer Indoor Flea Guard.

The most effective way to use methoprene is to monitor the flea population in your home and apply the spray when you notice an increase as the summer or fall weather arrives. Follow the directions on the label carefully. It will then prevent the exploding flea larvae population from maturing into adult biting fleas. The effectiveness should last 75-120 days, depending on the product used. The liquid form is longer lasting than the fogger, and it can be limited to only the appropriate areas in the home. Apply Precor only when necessary, using the white sock test to determine when to spray. If insect growth hormones are used very frequently over a long period of time, the fleas may be able to develop a resistance to them.

Although EPA tests have not indicated any potential long-term health problems with methoprene, it is always preferable to control the fleas with other less toxic methods whenever possible. However, if you need to use a pesticide spray, methoprene appears to be much safer than the more commonly available pesticides on the market.

7) Flea trap

There is one type of trap that may help to reduce the flea population in your home. You will have to judge its effectiveness for yourself, as no scientific studies have been done with it, although commercial varieties of the trap are now available. Fill a large shallow pan with soapy water and place it under a gooseneck lamp so the light shines directly on the water. The fleas are attracted to the light, and land in the water to drown. Turn off all the other lights in the room. Try to prevent pets and children from drinking the water.

In your yard

1) Safer® Indoor Flea Guard™

Safer Indoor Flea Guard can be safely used on areas outdoors that have a heavy infestation of fleas. Identify these areas by using the white sock test. Limit your spraying to only those locations that have a lot of fleas to avoid killing more beneficial insects than necessary. Direct the spray to the ground where the fleas live and breed. As the Flea Guard will only kill adult fleas, you may need to reapply the spray when the white sock test shows that more fleas have hatched.

If you will be using the soap frequently over large areas, it will be less expensive to buy Safer's® Insecticidal Soap Concentrate and dilute it yourself. A one to two percent soap solution should not damage most plants, especially if you spray during the cool morning or evening hours. If you want to be sure, you can test a limited area of any valuable plants. The Insecticidal Soap contains the same ingredients as the Flea Guard and is just as effective in killing the fleas.

Note: Safer's soaps will not work well if mixed with hard water. Make up a small batch of the solution in a jar and shake it. If no foam develops, you have hard water. In this case, buy a bottle of soft, de-ionized or distilled water to mix with these soaps.

2) Diatomaceous earth

During dry weather, diatomaceous earth can be useful to kill the fleas in your yard. This powder can kill both the flea larvae and the adults if it doesn't rain hard for several days after application. Again, be sure to only dust those areas that the white sock test shows to be heavily infested, in order to limit the number of beneficial insects killed by the diatomaceous earth. (See pages 57-59 for more information on the safe application and use of this powder.)

3) Pyrethrum

Although pyrethrum powder or spray can be used to kill the fleas inside or outside your home, try using the other suggested methods or products first. The powder is fairly expensive to use over large areas, and can cause strong allergic reactions in some people. Most natural pyrethrin sprays contain piperonyl butoxide (see the precautions on pages 63-65). Both the Flea Guard and diatomaceous earth are less toxic to use than pyrethrum, and the Flea Guard has a good quick-kill capability. Unfortunately, methoprene (Precor) biodegrades too rapidly outdoors to be of any use there.

Chapter 3

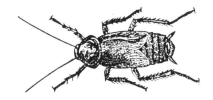

Cockroaches

Cockroaches have been around for millions of years, and have probably coexisted with humans for as long as we have had homes. There are many different species, with various sizes and coloring. The environments they prefer vary from warm to cool, and moist to dry. They breed incredibly fast all year.

Cockroaches can eat anything that we can, and many things that we can't (such as paper, soap, and shoe polish). They love dark, crowded environments where they feel surfaces touching them all around. Roaches have been able to learn to avoid locations that have been treated with chemical pesticides that they can smell. Like most other insects, they have also become immune to many pesticides.

Unfortunately, roaches are capable of transmitting a wide number of serious diseases, including hepatitis, salmonella, and parasitic toxoplasmosis. It's important that you throw away any food that the cockroaches have been able to get into. Saving a small amount of food is not worth the risk of an illness.

CONTROL METHODS

University studies have shown that the proper application of boric acid powder can provide a better than 99% control of roaches for more than three months. Many people have found that this alone provides adequate control.

In locations where it is not practical to use the boric acid, there are several alternative control methods available. These include the application of Gencor™ (an insect growth regulator), Diacide, or pyrethrum, and the use of homemade baited traps, commercial traps, and boric acid baits.

In addition, it is a good idea to clean out existing and potential roach nesting sites. To help prevent future infestations, you can close off the entrances that cockroaches use to enter your home. In some situations you may also find it helpful to reduce their access to food and water sources.

Boric acid powder

Boric acid is very effective for controlling cockroaches and other insects. It may take up to one week to kill the existing population, but the effects are long-lasting. The powder will continue killing roaches that enter your home for months. (For complete information on the safe and proper application of boric acid, see pages 60-62.)

In rooms where you find cockroaches (especially near potential food and water sources), dust the boric acid powder into the cracks along baseboards and moldings, underneath and behind stoves and refrigerators, around plumbing fixtures, and into openings in the walls. Be sure to squirt the powder into the crevices around sinks, countertops, and cupboards. Another favorite cockroach location is underneath the kitchen and bathroom cabinets.

Roaches can fit into a space only one-sixteenth of an inch high, so don't ignore tiny cracks.

Apply the powder so that it forms a fine layer on the surfaces. The roaches will just walk around any piles of the powder. Wipe up the boric acid which doesn't get into the cracks or holes. Remove your food, and eating and cooking utensils, before applying the powder in the kitchen. Don't bother using the boric acid in locations that are damp or wet, as the powder will only work when dry. In humid climates, you may need to reapply it every four to six months if the roaches start increasing in numbers again.

If you have time after applying the boric acid, reduce the hiding places that the roaches have access to by sealing the cracks with caulk. Caulk comes in a wide variety of forms (tube, rope, and caulking gun), so choose the type that you find easiest to use. Also be sure to glue down any loose wall paper, as it makes a wonderful nesting site for the roaches.

Other control methods

1) Homemade baited traps

In damp or wet locations where boric acid powder is not very effective, baited traps can be helpful. Completely wrap the outside of several glass jars with masking tape (so the roaches will be able to climb into them), and spread a thin two inch wide band of petroleum jelly just below the inside rims. Place a bait (pet food, banana peel, beer, or other food) in the jars, and put the traps upright in various corners of infested rooms. Empty the caught roaches and silverfish into a bucket of hot, soapy water. Commercial roach traps and boric acid baits are also available. As usual, keep all traps and baits out of the reach of pets and children.

2) Diacide

Diacide (a combination of diatomaceous earth and pyrethrin) is a very effective product against roaches. Several cities are successfully using Diacide for cockroaches in their sewers. Apply the powder in the same locations where you would use boric acid. (See page 65 for more information on this product.)

3) Gencor™

One product that is very effective for long-term control of roaches is Gencor, an insect growth regulator. (For an explanation of insect growth regulators, see page 26 in the flea chapter.) It will not kill adult roaches, but it will prevent the young ones from maturing and breeding. The spray is applied to the same locations where you would use boric acid. Make sure that it doesn't come in contact with your food, or with your eating or cooking utensils. You should also avoid contact with fabrics and upholstery. Follow the directions on the product label. The effect of the spray will last for several months. To control the adults, use traps, baits, Diacide, or pyrethrum powder.

Gencor is often combined with other more toxic pesticides, so read the label carefully. If you can't find a local source of a product with Gencor alone, check with wholesale suppliers of pest control products, or contact the Natural Gardening Research Center listed in the appendix.

Clean out potential nesting sites

Cockroaches can infest and nest in a wide variety of locations—such as televisions, control panels of microwave ovens, refrigerator insulation, electric clocks, boxes of old clothes, and piles of newspaper. Some species also like draperies, bookcases, water meter vaults, or dense vegetation.

As unpleasant as it might be to do, go through your home and try to identify the locations where the cockroaches are concentrated. As roaches are usually more active after dark, it may help to do this at night, with a narrow-beamed flashlight. Watch where the roaches go to hide, and find out where the largest populations are.

Throw out piles of grocery bags, newspapers, or corrugated boxes that are infested. Shake out and vacuum draperies or bookcases that are hiding roaches. Remove dense vegetation next to your home, and keep piles of wood or other debris at a distance. Rubber shower mats with suction cups should be hung up to dry after each use. Clean out radios, microwave ovens, or other items that you find the roaches nesting in. Luckily, many homes have only two or three major nesting sites, often in the kitchen area.

Close off entrances into your home

Cockroaches often enter homes by hitchhiking rides in grocery bags, cardboard boxes, beverage cartons, furniture, and appliances. If you buy used furniture or appliances, check them over very carefully for evidences of infestation. Empty your boxes and grocery bags as soon as they are brought in, and then dispose of them outside. Occasionally, cockroaches will come up from the sewer through your drain pipes into your sinks. If this is happening to you, the simplest solution is to keep a tight-fitting plug in the drain whenever you are not using the sink.

Warm Climates

Roaches can also enter the home along pipes and cracks in the outside walls, especially in areas with a warm climate. Seal the openings around the water and gas pipes that enter the outside walls, and use caulk to seal cracks around windows and doors.

Apartments

In apartments, roaches frequently enter from the surrounding units through the heating and air-conditioning vents, and along the pipes in the bathrooms and kitchens. To prevent this, cut pieces of fine window screen (nylon or wire) and tape them over all the air vents. Using plaster or caulk, seal all of the cracks around the water and gas pipes where they enter through the inside walls. Be sure to seal off any other crevices or holes that you can find. To keep the roaches from entering from the apartment building hallway, apply weather stripping around your front door.

Reduce the sources of food and water

While reducing the sources of food and water in your home can help to control the cockroach population, by itself it is not likely to eliminate them. Roaches can survive a month without water, and up to three months without food. However, in some situations you may find it very helpful to take the following steps:

Water

Fix dripping faucets and leaky pipes. Keep the countertop by the kitchen and bathroom sinks dry. Dry out your wet dishrag or sponge in the open, not under the sink. Cover the fish tank with a fine mesh screen, and dry up the drip pan under the refrigerator.

Food

If you can't wash the dishes after every meal, put them in a pan or sink full of soapy water. Keep your garbage in a tightly sealed plastic bag or container. If possible, take out the garbage every evening (before the roaches' happy hour). Clean the floors, stovetop and countertop regularly, especially before bedtime.

Store food in tightly sealed containers or in the refrigerator (including pet food). Clean up your pet's food bowl and eating area soon after each meal, or place the food bowl in a shallow pan of soapy water so the roaches don't have access to it.

Chapter 4

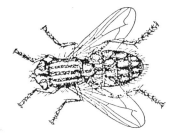

Flies

While most people consider flies to be a nuisance, not many realize how truly dangerous flies can be. Even in today's relatively sanitary society, flies are capable of transmitting a wide variety of mild and serious illnesses.

Flies can breed at an incredibly fast rate. In warm weather, the process may take only seven to ten days (from egg, to maggot, to pupa, to hatching as an adult). Adult female flies can lay hundreds of eggs in their three to four week lifespans. The maggot larvae can squeeze through tiny cracks and travel up to 50 feet to reach a food source. Flies will feed and breed on nearly any moist decomposing material. This includes household garbage, animal manure, fermented grass clippings, and dead animals. Flies like to congregate together, and they usually gather in the areas of brightest light.

Not all species of flies are dangerous. There are many harmless kinds that pollinate plants, destroy insect pests, or fill other niches in the ecosystem. With the exception of the cluster fly, most of these beneficial flies will not be found in your home.

Cluster flies breed mostly in clean, rich soil that has plenty of earthworms. They do not normally carry disease. You will usually notice them when they gather in your windows on warm winter days. This means that they have found a place in your home to hibernate over the winter (sometimes inside the walls or under the floor). The best way to control these flies is to locate the cracks that they are using to enter your home, both inside and outside, and seal them tightly with caulk. Also be sure to keep your window and door screens in good condition.

CONTROL METHODS

There are a wide variety of methods to control flies, both indoors and outdoors, depending upon the source of the problem.

Indoors

1) Screens

Use screens on all windows and doors, and keep them in good condition. The screen doors should be adjusted to close rapidly after opening.

2) Darkening the room

If a room has many flies in it, you can often greatly reduce the number by darkening the room. Close all of the curtains, with the exception of one door or window on the brightest side of the room. Open the window or door, and quietly shoo the flies toward the opening. They are naturally attracted to the brightest area. When you have chased out as many flies as possible, close the screen.

3) Fly paper

One old-fashioned method to control flies is the use of fly

paper. It is still effective and nontoxic, although many people don't appreciate having their hair or curtains stuck to the paper along with the flies. It unfortunately doesn't usually blend in with the decor of the room, either.

However, there are sticky tubes now available which are easier to handle. The fly paper opens up in the form of a rigid tube, and it often comes with flies already printed on it (to encourage real flies to stop by for a permanent visit). It may also include the use a fly sex attractant (a pheromone). As usual, be sure to keep the fly paper or tubes beyond the reach of children and pets.

4) Indoor fly trap

A newer alternative is the use of an indoor fly trap produced by Safer, Inc. The flies are attracted into it by the use of a pheromone (a sex attractant). When the flies die, they drop into a disposable bag. This fly trap is attractive and odorless. (Note: Most other fly traps use strong-smelling baits, and should only be used outdoors.) All traps should be kept out of the reach of pets and children.

5) Fly swatter

Don't forget the option of the calorie-burning hand-held fly swatter.

6) Pyrethrum

Pyrethrum can be used for temporary control of the flies. It biodegrades too quickly to offer long-term control. Pyrethrum should be your last choice method. Do not use it if you are able to control the flies by any of the other methods mentioned above. The least toxic way to use pyrethrum inside your home is to limit the spray to just the window areas where the flies congregate. (See pages 63-65 for more information on pyrethrum.)

Often, the best way to control flies inside your home is to control them outside of your home first.

Outdoors

Check for possible sites of fly breeding on your property. One garbage container can produce thousands of flies a week. Dog manure in the back yard or piles of fermenting grass clippings can also be sources of problems. If the flies are coming onto your property from somewhere else, fly traps are probably your best solution. (For more information on fly traps, see page 42.)

1) Garbage

Don't keep your garbage only in plastic bags. They are too easily torn by accident or by neighborhood animals. Metal garbage cans should be kept elevated off the ground so the bottoms can remain free of rust. The cans must have no holes or cracks, and the lids should seal tightly. Remember that flies can breed and mature in just seven days. In warm weather, consider arranging for twice-weekly garbage pickup.

2) Grass clippings and compost

When piles of grass clippings ferment, they produce an ammonia smell that is very attractive to flies. One solution is to break up the piles of grass so that they will dry out rather than ferment. The grass can then be used as a thin layer of mulch in the flower or vegetable garden, or it can be added to a compost pile. There are many excellent books available that describe a variety of ways to make good fly-free compost. Many of these are suitable for city or suburban settings. If you choose not to use your grass on your own gardens, there are probably many gardeners in your area that would be happy to take your grass for free. Talk to your neighbors or post signs at nearby stores.

Having your grass sent to the landfill should be your last resort.

3) Animal manure

Flies can breed as easily in dog droppings, as in horse or cow manure. There are several ways to effectively deal with this problem.

Diatomaceous earth (D.E.), when added to an animal's diet, can greatly reduce the number of flies that breed in the manure. For cats and dogs, the suggested rate is one-half to one teaspoon for small animals, and up to three teaspoons for large dogs. The D.E. should be given every day, thoroughly mixed with the pet's food. D.E. can be fed free choice to larger animals if it is kept in a bucket or other container out of the rain. It can also be blended in with their food at a rate of two to three percent of the dry weight of their total food intake. (See pages 57-59 for more information on the safe and effective use of diatomaceous earth.)

If you are not using fly parasites, diatomaceous earth can also be regularly and liberally dusted in animal stalls and pens, dog runs, corrals, on manure piles, and any other fly breeding locations.

Fly parasites are very useful when you have larger numbers of animals to deal with, as in dog kennels, horse stables, and livestock barns. The parasites (usually a mixture of several different species) kill the flies in the pupal stage. They should be released bi-weekly or monthly throughout the fly-breeding season. When combined with basic water and manure management, they can reduce the adult fly population by over 90%.

The parasites work best when they are first released in the spring as soon as adult flies start appearing. If significant fly populations have already become established, larger initial releases of the parasites will be required. To bring the adult flies under control quickly, this will need to be combined with the use of large outdoor fly traps or with the selective use of pyrethrin sprays. Listed in the appendix of this book are suppliers of fly parasites. Contact them to obtain the best control recommendations for your particular situation.

Pyrethrin should never be sprayed near or on the manure, as it will also kill the helpful parasites along with the flies. Find the sunny walls, fences, and other locations where flies are congregating, and limit the spraying to those areas. It should only be used temporarily until the flies are brought under control with other methods. (For more information on the safe and effective use of pyrethrin, see pages 63-65.)

Diacide (a combination of diatomaceous earth and pyrethrin) has been used very effectively for many years to control flies at stables, livestock barns, and county fairs. Using electrostatic dusters, the powder is applied onto and into the buildings, where it forms a fine coating on all the surfaces. Insects landing on these surfaces are killed. The combination of diatomaceous earth and pyrethrin sustains the killing capability of the pyrethrin over a period of several days instead of the normal several hours. (For more information on this product, see page 65.) Be aware that even biodegradable pesticides, such as pyrethrum, can disrupt the local ecosystem. When you are able to bring the flies under control using the other less toxic methods (e.g. parasites or fly traps), discontinue using Diacide.

4) Fly traps

Fly traps are most helpful when combined with other control measures, but they are sometimes very effective when used on their own. When the source of the flies is not on your own property, the traps become almost necessary. There are a wide variety of types and sizes of fly traps available, and there are several different baits to choose from.

Smaller traps are best suited for backyards, while the medium to extra large traps work well for farm and stable locations. The traps should be located near fly breeding or congregating areas. If flies are coming into your area from elsewhere, place the traps on the edges of your property. Fly traps usually work best at ground level, but it is more important to keep them out of the reach of pets and children. Most baits are strong smelling, so you should place the traps at a comfortable distance from your home.

Effective fly baits range from pieces of meat, fruit, molasses (three parts water, one part molasses, one package active dry yeast), to commercial mixes of ammonium carbonate and yeast. The ammonium carbonate and yeast mixture is one of the most commonly used. All baits must be kept constantly moist, as dry baits will not attract flies. Meat or fruit can be kept in a shallow layer of water. Most other mixes are water-soluble.

The instructions with the trap should tell you which general species of flies it was designed to catch. If you are not sure what kind of flies are bothering you, the local cooperative extension service should be able to help you identify them. You may occasionally need to experiment with different types of baits and traps to find one that is most effective for your particular situation.

5) Miscellaneous information

There are large sheets of fly paper available for use in barns and stables. They should be hung near where the flies congregate, but kept out of the wind and sun. You should be aware that this paper can catch beneficial insects, small bats, and hummingbirds. Check it regularly to see if this becomes a problem.

Do not use outdoor electric zapper lights. They are not very effective against flies, mosquitoes, or other insect pests, and they have been proven to kill large numbers of beneficial insects.

Be very careful about using poison fly baits available at feed and grain stores. Many of them contain highly toxic materials that should be avoided.

Chapter 5

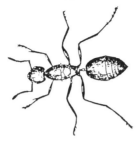

Ants

Ants live in colonies made up of several hundred to several thousand sterile worker ants, a few fertile males, and one or more queens (egg-laying females). These colonies can be located in the soil, under rocks and pavement, in trees, or in buildings. The workers search for food to bring back to feed the colony. The type of food they prefer (sweets, meat, starch, or grease) depends on the species and the time of year.

While it's a good idea to keep your home free of ants, keep in mind that ants actually help us out a great deal in the environment. They clean up dead insect and animal bodies, which helps to control flies. Ants also prey on fly larvae, cockroaches, and a wide variety of other pests. They eat the eggs of many insects, including mosquitoes. Ants also aerate and loosen the soil, encouraging plant life, and helping the soil to absorb rain faster.

CONTROL METHODS

If ants are bothering you in your home, there are several methods of control available. The safest methods involve removing the sources of food that are attracting them, and sealing off the ants' entrances into your home. If this doesn't work, you can then try killing the ant colony by the use of boiling water, diatomaceous earth, pyrethrum, Diacide, or boric acid baits.

Remove the source of attraction (food)

Clean up the sources of food for ants in and around your home. Your garbage should be kept in tightly sealed bags or garbage cans. Keep kitchens and other rooms as free of crumbs and spills as possible. If you can't do your dishes after every meal, put them in a sink full of soapy water. Store food in tightly sealed containers and wipe down the outsides of honey and jam jars.

Even pet food (dry or moist) can be a source of problems. It should be kept in tightly sealed bags or containers. Clean up the food bowl and eating area soon after your pet is done eating. If you must leave the food out for long periods of time, block the ants' access to it by placing the food bowl in a slightly wider shallow pan of soapy water.

Block their entrances into your home

Look for the cracks and the holes in your home (both indoors and outdoors) that the ants may use to enter. Indoors, this is usually fairly easy—just follow the lines of ants back to their source of entry. Dust boric acid into the openings, and then seal the cracks tightly with caulk, plaster, or (for quick, temporary results) petroleum jelly. If they are coming in under a door, apply good weather stripping.

The ants may locate another access hole. If they do, repeat the procedure. Be sure to vacuum up or kill off any stragglers in your home. Dispose of the vacuum bag outside or suck some pyrethrum dust or Diacide into the vacuum to kill them. (For complete instructions on the use of boric acid, see pages 60-62.)

Kill off the colony

If cleaning up food sources and sealing off entrances into your home isn't effective, there are several ways to kill off the colony. You may find the nest site in the ground next to the house or many feet away. It may even be in the walls or under the floor. Locating the colony will take some close observation and guesswork. Just follow the ants back to the holes they are using. In some cases you may never find it.

1) Boiling water

You can sometimes trace the ants back all the way to their colony outside. If the nest is located in open ground or under rocks that you can move, you can pour large amounts of boiling water into the nest site. Do not use cold water. It will not kill them, and may encourage them to look for a higher, drier location such as your home.

2) Diatomaceous earth

During dry weather, you can use diatomaceous earth. This is also helpful to use when the colony is located under sidewalks or pavement where it is difficult to thoroughly saturate the nest with boiling water. Spread a generous quantity over the nest entrances and for several feet in all directions. Reapply it after a hard rain. Plain diatomaceous earth can take one or two weeks to kill insects. (For more information on diatomaceous earth, see pages 57-59.) Do not use boric acid in this manner where it will be exposed to animals and children.

3) Pyrethrum

Another way to kill the ants is to use pyrethrum, either as a powder or as a spray. It offers a quick-kill capability that biodegrades rapidly. Use the pyrethrum around the entrances to the nest site and along the ants' pathways. You may need to apply it several times to kill the entire colony. (For more information on the safe use of pyrethrum, see pages 63-65.)

4) Diacide

An even more effective product is Diacide, which combines diatomaceous earth and pyrethrin. The pyrethrin is absorbed into the diatomaceous earth, sustaining the killing capability of the pyrethrin over several days. The diatomaceous earth also provides long-term control. (See page 65 for more information on this product.)

5) Boric acid baits

Poison baits are a common ant control method. Many of the poisons available are very toxic and long lasting. Baits containing boric acid (or borax) are among the least toxic available. The boric acid is usually mixed with a sweet or greasy bait. You may need to experiment to determine which one is attractive to the ants presently invading your home. While it is possible to make your own bait, it is safer to use a commercial bait that comes in a container that is difficult for children and pets to open.

It is important to remove all other food sources available to the ants so they will make full use of your bait. Observe the ants to be sure they are taking it back to the colony. These baits work best if they are placed along the ants' trails or at the entry points to your home. While it may take several weeks to completely wipe out a colony, you should notice a reduction in the population within a few hours or days.

Some of these baits are sold as a paste, gel, or liquid. Others come in various types of containers. No matter what kind of bait you use, make sure that it is <u>always</u> out of the reach of children and pets.

Chapter 6

Pantry Pests

There are a number of insects that will infest food stored in your kitchen and pantry. These include the larvae of meal moths, and various beetles, weevils, and mites. They can come into your home from outdoors, or, more likely, they will hitch a ride in your shopping bags. These pests can eat anything from flour and grains to pasta, dried fruit, legumes, cheese, meat, and even herbs and spices.

Probably the most commonly noticed pests are the meal moths that infest flour, grain, dried fruit, and nuts. You may notice lumps of food clinging to the sides of a container, or find larvae or cocoons in your food. Sometimes the first thing you see are the adult moths flying away when you open a container.

Prevention is the best way to avoid these pests. Inspect all containers of dry food (including dry pet food) for very tiny holes or loose cardboard flaps. If a package is leaking a small amount of flour or mix, it may already be contaminated. If you notice a package is damaged after you have brought it home, take it back to the store immediately. Don't take a chance on having the insects spread to your other food.

Buy only moderate quantities of food, and use up the old stock before you use the new. Keep your pantry shelves clean and dry. Once you have opened a plastic or cardboard package, store that food in tightly sealed containers (of glass, metal, or sturdy plastic) to prevent the insects from gaining access to it. Don't combine old foods with the new.

Seal any cracks in your kitchen or pantry walls, floor, or ceiling. Check your stored foods regularly for infestation (at least weekly during warm weather). Throw out any food that looks contaminated. Try to keep your pantry cool and dry. Keeping vulnerable food in the refrigerator or freezer can help to protect it. Many food pests die after prolonged freezing. Storing dry food in a cold freezer for a few weeks will kill off nearly all the pests, and even a few days can greatly reduce their numbers.

These pests can also be killed by heat. Spread the food thinly on cookie sheets, and heat it to about 125° F for two hours. Don't let the temperature get much above that. Cool the food completely before repackaging it. Infested dried fruit can be dropped into boiling water for one minute. Dry the the fruit thoroughly before storing it in a tightly sealed container.

Large amounts of grain can be protected by thoroughly mixing it with diatomaceous earth. (See pages 57-59 for more information on this product.) Use one cup of food-grade diatomaceous earth for every 25 pounds of grain. The grain will appear dusty, but it does not need to be cleaned off before being used. Again, store the food in a tightly sealed container. This will help to keep moisture out of it.

Another method to help protect your food is to use pheromone (sex hormone attractant) traps to catch adult moths before they breed and lay eggs in your food. (Suppliers of these traps are listed in the appendix.)

Chapter 7

Clothes Moths

The larvae of clothes moths like to eat items made of wool, hair, or feathers. This includes wool clothes and blankets, down quilts and pillows, piano felts and old felt insulation, and infrequently used stuffed toys and vacuum cleaner bags. They may also be attracted into your house, basement, garage, or attic by dead rodents, birds, and insects.

Keep your window and door screens in good condition, and seal any cracks or crevices around your baseboards, doors, and windows to help keep clothes moths out. Also try to rodent-proof your home. Check your unused and stored items regularly for infestations. Vacuum the felts in your piano frequently.

Frequently used items are safe from moth damage. However, items that need to be stored for a few weeks or months should be carefully prepared. Wash or dry clean fabrics before storing them. Another alternative is to hang the item outside in the sun for a day and thoroughly brush it down to remove eggs and larvae. Be sure to pay special attention to pockets, collars, pleats, and seams. If you can't wash, dry clean, or sun an item, you can put it into a cold freezer for a few days to kill the insects.

Once you are sure the items are free of pests, you can safely store them in any clean, airtight container. They can be stored in heavy plastic bags or cardboard boxes, as long as all of the seams and holes in these containers are sealed tight with tape. You can then safely store the items in a cool, dry location. A spare closet or room is generally better than the attic or basement.

Herbs have been used for centuries to repel moths. It is best to prepare and store your items according to the previous instructions, and to use herbs only as a supplementary measure. You can also use herbal satchets in a drawer with woolen items that you use only occasionally, but don't want to store away in a box.

Cedar chests can be useful, but they need to be airtight. They should be renewed yearly by either lightly sanding the cedar wood inside the chest to release more scent, or by applying fresh cedar oil. Other helpful herbs include bay leaves, lavender, mint, tansy, southernwood, citrus peel, and many others. Crush or squeeze these herbs often to release more scent.

Termites

Termites are the one insect pest that will most likely need to be exterminated by a professional. They can cause extensive damage to your home and property. There are nonchemical treatments available, although they do have some limitations. One method is the use of a disease microorganism, called Spear, that kills underground termites. Another safe treatment method is the Electrogun. It kills drywood termites by electrocution. To obtain up-to-date information on safe exterminating methods for termites, contact one of the organizations listed on page 73. They may also be able to suggest methods for preventing termite infestations.

Spiders

While it is unlikely that you will ever be able to completely eradicate spiders from your home, there are several methods that you can use to keep their numbers to a minimum. Seal the cracks that they can use to enter your home—around baseboards, doors, windows, walls, and ceilings. Keep your window and door screens in good condition, and apply tight weather stripping around them. Clean up any clutter around your home to reduce potential hiding places. Vacuum your home thoroughly and frequently, especially in corners, under furniture, and behind curtains and appliances. Clear out any vegetation around the base of your home. Spiders are often more resistant to pesticides than many other insects. If you have a serious problem with black widows or other dangerous spiders around your home, you can try dusting Diacide once a week around the base of your home, under your porches, in basement windows, or other locations that they frequent. (For more information about this product, see pages 63-65.)

Chapter 9

Diatomaceous Earth

Diatomaceous earth (pronounced die-uh-toe-may-she-us) is a finely ground powder made from fossilized single-cell algae known as diatoms. The diatoms lived in huge numbers in the oceans millions of years ago. As they died, they built up large deposits of their silicon shells on the ocean floor. These deposits (those now on dry land) are mined, ground, and screened to produce a very safe, nontoxic pesticide.

The diatom shells are made primarily of hard silicon, and they have microscopically sharp edges when properly ground. These sharp edges can pierce an insect's exoskeleton and cause it to die of dehydration. If the insect swallows the dust when cleaning itself, it will be damaged internally. The diatom shells are also capable of absorbing fluids, including the wax covering an insect's body. This also encourages dehydration.

The internal and external tissues of mammals are very resistant to the tiny edges of the diatom shells, so it is very safe for people and their pets to handle and even swallow the powder. The main precautions necessary are to keep the powder out of the eyes, and to avoid breathing large quantities of it. In addition, prolonged exposure to the powder may dry out the natural oils on your skin.

Some insects are more resistant to damage than others, but the powder has been used fairly effectively against roaches, fleas, ants, flies, and other household pests.

Because diatomaceous earth kills insects by dehydration, it can take up to a week or two for the insects to die. If the humidity level in the air is very high, the insects may not dehydrate at all until the air gets drier.

It is very important to only use food-grade, natural diatomaceous earth, and not the swimming pool grade. The diatomaceous earth used in swimming pool filters has been both chemically treated and heat treated, and is extremely hazardous to breathe in or swallow.

Natural diatomaceous earth is composed of non-crystalline (amorphous) silica, and it can apparently be dissolved by the human body. This makes it much safer to use. Try to make sure that the brand of natural diatomaceous earth you buy has a less than 3% crystalline silica content. (Swimming pool grade has about 60-70%.) One brand, Perma-Guard® (produced by Universal Diatoms, Inc.), has less than 1% crystalline silica. There may be some brands on the market with as much as 7%, which is considered dangerous by the World Health Organization. Ask your supplier if you're not sure your brand is safe.

Diatomaceous earth has been approved by the EPA to be used as an insecticide in stored grain. It will kill the insects that would otherwise feed on the grain. Diatomaceous earth does not affect the baking abilities of the grain when it is ground into flour, and it is completely nontoxic to anyone eating it. Farmers using it regularly in their animals' feed have found it to be an effective wormer for their animals, and have noticed a large reduction in the number of flies breeding in their animals' manure.

Diatomaceous earth has been used to kill fleas in the home and yard. (See pages 25-26 for instructions on how to do this.) A simple dust mask should be used when applying the powder indoors. The powder is an extremely fine, inert material, and will not damage rugs or floors.

Your rooms can be occupied as soon as you lightly vacuum the dust from the surfaces. When properly applied, the home will not look or feel dusty.

Universal Diatoms, Inc. also produces another diatomaceous earth product called Diacide. It adds natural pyrethrin and piperonyl butoxide to the powder to create a pesticide that is more effective than the original ingredients are on their own. (For more information, see pages 63-65.)

Chapter 10

Boric Acid

Boric acid can be very effective in controlling ants, roaches, and silverfish in your home. It is odorless and doesn't give off fumes, nor does it produce cancer, birth defects, or nerve damage. However, like many other common household products, it can be harmful if swallowed. One tablespoon will be toxic to a young child.

Follow the instructions on the product label. Always make sure that all containers are clearly and permanently labelled. Keep them out of the reach of children and pets.

Properly used, boric acid is safer and often more effective than most of the chemical pesticides on the market today. Boric acid should be placed in cracks and crevices where the insects will come into contact with it, and where people and pets won't be exposed to it. Applied in this manner, it would be difficult for anyone to come into contact with enough boric acid to become ill by eating it.

Boric acid kills insects in two ways. As a dust, it clings to their bodies when they walk through it and penetrates the waxy coating on their exoskeletons. They also will swallow the poison when they clean themselves. It may take up to a week to kill most of the pests, but boric acid should remain effective for several months.

Most insects are able to develop an immunity to frequently used pesticides. Boric acid, however, has proved to be an exception so far. Roaches have not become immune in several decades of use.

Buy only 99% technical boric acid powder. It has been tinted a color (generally green or blue) so that it can't be confused with flour, sugar, or other food. In addition, it has been electrostatically charged. This prevents it from clumping and allows it to spread as a fine layer of dust on surfaces. It also helps the powder to cling to the insects' bodies when they walk over it.

Do not use medicinal boric acid. It is not nearly as effective, and it is too easily confused with other substances.

One excellent product is Roach Prufe®. There are also many other brands of boric acid on the market. Some come prepackaged in squeezable bottles for dusting. Don't buy boric acid in a granular form. It doesn't cling very well to the insects' bodies.

To apply the powder, you can use a squeezable plastic ketchup bottle. You can also use any small commercial duster with a pointed nozzle that you can aim into cracks. Wear a disposable dust mask and gloves when applying it to a large area. The powder is slippery, so be sure to clean it off the floor areas where you walk. Always clean up any boric acid that has not gotten into the cracks and crevices so that children or animals cannot reach it. Use a slightly damp rag. If the rag is too wet, it could get the powder in the crack too damp to be effective. Avoid dusting your houseplants or their soil with the boric acid, as it could damage them.

Kept dry and undisturbed, boric acid remains effective for well over a year. In humid climates, however, you may need to reapply the boric acid after four to six months if the insects reinfest your home.

Boric acid is also used in baits. These baits are likely to

be more effective than the powder in very damp or wet locations. It works more slowly than many other insecticides, and it can take anywhere from a few hours to several days to significantly reduce the insect population. This is actually very helpful when it is used in ant baits, as the worker ants have a chance to feed most of their colony with the poison before it starts killing them. With fast-acting poisons, the ants may stop taking the bait before they feed much of the colony.

It is not a good idea to make dough balls or other types of flour baits that are laced with boric acid. They are too tempting to pets and children. Instead, use commercial baits available in containers with holes too small for curious fingers. In any case, <u>always be sure to keep any bait out of the reach of children and pets</u>.

Chapter 11

Pyrethrum

Pyrethrum powder is a natural product made from dried, ground up pyrethrum flowers, daisy-like members of the chrysanthemum family. The active ingredient in this powder is called pyrethrin. Pyrethrin is a mild, contact nerve poison. It will kill ants, fleas, ticks, flies, silverfish, spiders, cockroaches, and a wide variety of other insect pests. It has a very low toxicity for warm-blooded animals, which are able to quickly metabolize the pyrethrin out of their bodies.

Pyrethrin biodegrades in less than one day in the environment, so it will only kill insects that come into direct contact with it soon after application. However, if pyrethrum is stored in a closed container in a cool, dry, dark location, it can remain effective and usable for up to three years.

A small percentage of people may have an allergy, sometimes with strong reactions, to pyrethrum. People that are already allergic to ragweed appear to be at higher risk. Many pyrethrin product labels encourage people to limit their exposure to the product in order to reduce the possibility of allergic reactions.

In over one hundred years of use, insects have not been able to become resistant to pyrethrum, unlike most chemical pesticides. However, because it is a mild poison, it does not often kill 100% of the insects that come into contact with it.

Pyrethrin often first causes the insects to become agitated and active. It then paralyzes their nervous systems, and they become stunned. If they receive less than a fatal dose, they may fully recover from the effects of the poison. Because of this, almost all of the pyrethrin products on the market also contain the chemical called piperonyl butoxide. It is not a poison in itself, but it does strongly increase the killing capability of the pyrethrin by slowing down the insect's ability to detoxify the poison, giving the poison a better chance to kill the insect.

Concern about the possible effects of piperonyl butoxide on people and pets is such that it is no longer allowed under the California Organic Foods Act. A few studies have indicated that piperonyl butoxide may increase the carcinogenic effects of other chemicals to which the animal is exposed. Considering the large number of chemicals that we are all exposed to every day, it may be advisable to avoid prolonged exposure to piperonyl butoxide.

If you choose to use a product with piperonyl butoxide, try to avoid inhaling it or leaving it on your skin for more than a short period of time. It should be safe to use if it is applied in limited areas (such as on ant nests), or in areas where people will not be in much direct contact with it (in cracks for roaches, or on barn walls for flies).

Pyrethrin is available in many products, but always check the label carefully. The manufacturers often add more toxic pesticides to the pyrethrin.

Natural pyrethrum powder (without piperonyl butoxide) is available from several sources. The most commonly available products are ZAP pyrethrum insect powder, and POW herbal flea powder, both produced by EcoSafe Products, Inc.

There is at least one pyrethrin spray available (made by Safer, Inc.) that does not contain piperonyl butoxide. It is called Safer® Yard & Garden Insect Killer. They have patented a formula combining the ingredients of their insecticidal soap and natural pyrethrin. It is both very effective and completely biodegradable. (Do not confuse this product with Safer® Garden Insect Killer, which contains pyrethrin and piperonyl butoxide.)

Another pyrethrin product is Diacide, produced by Universal Diatoms, Inc. It combines diatomaceous earth with pyrethrin and piperonyl butoxide. The advantage of this product is that the pyrethrin is absorbed into the diatomaceous earth to produce a sustained-release effect. The pyrethrin will continue to kill insects on contact over a period of several days. The diatomaceous earth can kill the insects by dehydration over an even longer period of time. Thus it is a very useful product, despite the piperonyl butoxide. A number of cities have successfully used it to kill cockroaches in their sewer systems. It has also been used for many years to control flies at large county fairs.

Researchers have recently developed many synthetic versions of pyrethrin. These are called pyrethroids. They are often long-lasting poisons, and some of them appear to be carcinogenic. If you feel you must use a pyrethroid, contact one of the organizations listed on page 73 to find the safest ones to use.

Appendix

Please be aware that the information in this appendix may change at any time. Companies sometimes move, and they often change the items that they sell.

Although this appendix has been written primarily for pest control inside the home, many of these suppliers also offer safe and effective products for the yard and garden. Some of the products mentioned below may also be available locally at garden centers or hardware stores.

Harmony Farm Supply
P.O. Box 451
Graton, CA 95444
(707) 823-9125
Catalog $2 (Refundable)

Supplies: Fly parasites, natural pyrethrum dust, several pyrethrum sprays (with piperonyl butoxide), diatomaceous earth, Roach Prufe® boric acid, liquid ant killer, pheromone trap for pantry grain moths, Safer® Housefly trap, outdoor fly traps, Safer's® Flea Soap and Indoor Flea Guard™, herbal flea collar, Spritz Coat Enhancer Spray, and an extremely wide variety of many other products and books for the organic gardener and farmer.

Natural Gardening Research Center
Highway 48
P.O. Box 149
Sunman, IN 47041
(812) 623-3800
Catalog free

Supplies: Roach bait stations with boric acid (also good for ants and silverfish), liquid ant killer, diatomaceous earth, outdoor fly trap, natural pyrethrum powders, pheromone traps for pantry grain pests, Precor™ 1%, Gencor™ 9%, Safer's® Indoor Flea Guard™, herbal flea collar and shampoo, Natural Animal™ Bits and Spritz Coat Enhancer Spray, Vitamin D rodent bait, and a very nice variety of other products for the organic gardener.

I.F.M. (Integrated Fertility Management)
333 B Ohme Gardens Road
Wenatchee, WA 98801
(800) 332-3179
Free catalog

Supplies: Diatomaceous earth, Safer's® Flea Soap and Indoor Flea Guard™, pheromone trap for pantry grain moths, citrus oil shampoos, fly parasites, and a wide variety of other products for the organic gardener and farmer.

Peaceful Valley Farm Supply
P.O. Box 2209
Grass Valley, CA 95945
(916) 272-4769
Catalog $2 (Refundable)

Supplies: Fly parasites, natural pyrethrum powder, pyrethrin sprays (with piperonyl butoxide), diatomaceous earth, Drax® Ant Control baits (for sweet or grease feeding ants), Roach Away boric acid, roach glue traps, Vitamin D rodent bait, Safer® Housefly trap, outdoor fly traps, pheromone traps for pantry grain moths, Natural Animal™ Bits and Spritz Coat Enhancer Spray, nutritional and herbal food supplements for pets, herbal insect repellants and shampoos, Safer's® Flea Soap and Indoor Flea Guard™, and an extremely wide variety of other products for the organic farmer and gardener.

Nitron Industries, Inc.
4605 Johnson Road
P.O. Box 1447
Fayetteville, AR   72702
(800) 835-0123 to order their free catalog
(501) 750-1777 for technical assistance and other questions

Supplies: Diatomaceous earth, Diacide, herbal powder, shampoo, and flea collar, and a variety of other products for the home owner and organic gardener.

EcoSafe Products, Inc.
P.O. Box 1177
St. Augustine, FL   32085
(800) 274-7387
Catalog free

Supplies: Natural Animal™ pet care line — natural pyrethrum powders, Spritz Coat Enhancer Spray, diatomaceous earth, Yeast & Garlic Bits, herbal flea repellants and shampoo, and other safe products for your pets.

Bountiful Gardens
19550 Walker Road
Willits, CA   95490
Catalog free

Supplies: Diatomaceous earth, natural pyrethrum powders, fly parasites, outdoor fly trap, Safer® Indoor Flea Guard™, and a very nice variety of seeds, books, and other supplies for the organic gardener.

Ohio Earth Food, Inc.
13737 Duquette Ave., N.E.
Hartville, OH 44632
(216) 877-9356
Catalog free

Supplies: Diatomaceous earth, Safer's® Flea Soap and Indoor Flea Guard™, and a number of other products for the organic gardener.

Breeders Equipment Company
Box 177
Flourtown, PA 19031
Information free

Supplies: Fleamaster® Flea comb, New Bequipco Flea Trap

Spalding Laboratories
760 Printz Road
Arroyo Grande, CA 93420
(805) 489-5946
Information free

Supplies: Fly Predators® - biological fly control

Beneficial Biosystems
P.O. Box 8461
Emeryville, CA 94662
(415) 655-3928
Information free

Supplies: D. FLEA KIT (includes complete instructions, 5 lbs. diatomaceous earth, dust applicator, dust mask, herbal flea collar, free telephone consultation), also sells pure pyrethrum powder separately.

Copper Brite, Inc.
1482 E. Valley Road, Suite 29
Santa Barbara, CA 93108-1241
(805) 565-1566
Information free

Supplies: Roach Prufe® boric acid (available in ACE or True Value Hardware stores, or contact Copper Brite to locate distributors in your area).

Pristine Products
2311 E. Indian School Road
Phoenix, AZ 85016
(602) 955-7031

Supplies: Diacide, diatomaceous earth.

Oakmont Investment Co., Inc.
44 Oak Street
Newton Upper Falls, MA 02164
(617) 449-1580
(800) 447-2229
Information free

Supplies: Complete line of Safer® insect control products — including their Flea Soap, Indoor Flea Guard™, Housefly trap, Insecticidal Soap, Yard & Garden Insect Killer, and many others.

Meadowbrook Herb Garden
Route 138
Wyoming, RI 02898
(401) 539-7603
Catalog $1, first class mailing

Supplies: Diatomaceous earth, herbal pest control for animals, organic pesticides for home and garden, and books on pest identification and control.

J.L. Price Products, Inc.
16316 W. Glendale Dr.
New Berlin, WI   53151
(414) 796-8270
Information free

Supplies: Big Stinky and Wee Stinky outdoor fly traps.

Necessary Trading Company
New Castle, VA   24127
(703) 864-5103
Catalog $2

Supplies: Fly parasites, diatomaceous earth, outdoor fly trap, Vitamin D rodent bait, pheromone traps for pantry food pests, and a wide variety of other products for the organic gardener and farmer.

Safer, Inc.
189 Well Ave.
Newton, MA   02159
(800) 544-4453   Free brochure about products.
Does not sell by mail order.

Supplies: Safer's® Flea Soap, Indoor Flea Guard™, Safer® Housefly Trap, Safer® Yard & Garden Insect Killer, and a variety of additional safe pest control products. Now available in many local gardening centers and other stores, or through other suppliers in this appendix.

Nichols Garden Nursery, Inc.
1190 North Pacific Highway
Albany, OR   97321
(503) 928-9280
Catalog free

Supplies: Natural pyrethrum powder, Safer's® Flea Soap, and other alternative products.

Ringer Natural Lawn & Garden Products
9959 Valley View Road
Eden Prairie, MN 55344-3585
(800) 654-1047
Catalog free

Supplies: Natural pyrethrin sprays (with piperonyl butoxide), diatomaceous earth, and a variety of other products for the organic gardener and homeowner.

Waterbury Companies, Inc.
100 Calhoun St.
P.O. Box 640
Independence, LA 70443

Supplies: Manufacturers of DRAX® Ant Kil Gel, DRAX PF®, MOP UP, and Borid—all insect control products made with boric acid. Contact Waterbury Companies for information on these products. They do not sell these products by mail order, only directly to distributors.

FOR MORE INFORMATION

The following two organizations offer up-to-date information on safe and effective methods of pest control, both indoors and outdoors. They both have a list of publications that they sell. If you write to them for an answer to a particular problem, please enclose $1 or $2 to help cover their costs of providing the information.

National Coalition Against the Misuse of Pesticides (NCAMP)
530 7th Street, S.E.
Washington, D.C. 20003
(202) 543-5450

The Bio-Integral Resource Center (BIRC)
P.O. Box 7414
Berkeley, CA 94707
(415) 524-2567

Bibliography

Bio-Integral Resource Center. "A New Tool for Managing Fleas." Common Sense Pest Control Quarterly vol. 1:1 (Fall 1984), p. 20.

Bio-Integral Resource Center. "Two Safe New Products For Managing Fleas." Common Sense Pest Control Quarterly vol. 2:1 (Winter 1986), p. 18- 20.

Bio-Integral Resource Center. "Vacuuming Helps Reduce Developing Fleas." Common Sense Pest Control Quarterly vol. 4:3 (Summer 1988), p. 18.

"Can You Control Ants?" Sunset Magazine vol. 180 (June 1980), p. 186A(3).

"chemicalWATCH Factsheet: Methoprene." Revised and reprinted from Pesticides and You vol. 6:3 (August 1986), page unknown.

Downs, Diane and Pacher, Sara. "Natural Flea Control; Get the Jump on These Pet Pests Safely." Mother Earth News (July-August 1987), p. 29-30.

Hunter, Linda Mason. The Healthy Home : An Attic-to-Basement Guide to Toxin-Free Living. Emmaus, Pennsylvania : Rodale Press, 1989. 313 pp.

Lifton, Bernice. Bug Busters: Getting Rid of Household Pests Without Dangerous Chemicals. New York: McGraw-Hill Book Company, 1985.   270 pp.

Marder, Amy. "Flea-Fighting Strategies." Prevention vol. 40 (June 1988), p. 81-83.

Marrs, L. Do You Love Cockroaches? Publisher unknown: John Muir Institute Inc., 1981. 14 pp.

Martin, Kent. "Better Roach Control." Organic Gardening vol. 32 (August 1985), p. 92.

Olkowski, William, Helga Olkowski, and Shelia Daar. "IPM for the Cat Flea." IPM Practitioner vol. 5:9 (September 1983), p. 7-11.

Pitcairn, Richard H. "Flea control from the ground up." Prevention vol. 38 (September 1986), p. 97- 101.

Pitcairn, Richard H. "The Great Flea Dilemma." Prevention vol. 38 (August 1986), p. 93-98.

Savage, Eldon P. et al. National Household Pesticide Usage Study, 1976-1977. Fort Collins, CO: Epidemiologic Pesticide Studies Center, Colorado State University; and the Environmental Protection Agency, November 1979.

Schultz, Warren, George Abraham, and Katy Abraham. "Flea Killers You Can Live With." Organic Gardening vol. 32 (July 1985), p. 823-825.

Segnar, K.E. "The War on Fleas." Cat Fancy vol. 32:6 (June 1989), p. 50-58.

"Winning the Battle Against Those Wiley Flies." Sunset Magazine vol. 179 (July 1987), p. 72-73.

Yepsen, Roger B., ed. "Repellents and Poisons: Natural Sprays and Dusts." Chapter 10 in Organic Plant Protection. Emmaus, Pennsylvania: Rodale Press, Inc., 1976. p. 111-128.

Index

P.S.

I welcome letters and comments from you about your experiences with the various methods and products I have mentioned, or with any others that you may have tried. What has worked for you? What methods have failed, or have caused bad reactions for your family or pets? I would also welcome your suggestions for additional information that you would like to see included in the next edition of this book.

I promise to read the letters that I receive from you, but I am sorry that I cannot answer your questions by mail, or correspond with readers on an individual basis. I will share much of the information that I receive from you with other readers as I update this book with information on new products and research.

I wish you the best of luck with your safe pest control efforts!

Debra Graff
P.O. Box 1316
Sterling, VA 22170

LIBRARY,
ST. LOUIS COMMUNITY COLLEGE
AT FLORISSANT VALLEY

BOOK ORDERING INFORMATION

Individual copies of this book
are available for $4.95 each,
plus $1 shipping.
(Virginia residents add 22 cents sales tax.)

Send a check or money order
payable to:

Earth Stewardship Press
P.O. Box 1316
Sterling, VA 22170

Please clearly print the name and address
that you want the book shipped to.

Bulk rate book prices available upon request.

Write for information on how this book
can be used for fundraising for
your organization.